QUANTUM COMPUTERS FOR USE IN PREDICTIVE ANALYTICS

by

Jacob J. Turan

In Partial Fulfillment of the

Requirement for the Degree

of

Master of Arts, Intelligence Studies

March 2016

American Public University System

Charles Town, WV

The author hereby grants the American Public University System the right to display these contents for educational purposes.

The author assumes total responsibility for meeting the requirements set by United States copyright law for the inclusion of any materials that are not the author's creation or in the public domain.

© Copyright 2016 by Jacob John Turan

All rights reserved.

Dedication

I dedicate this thesis to both of my brothers, for sparking and keeping alive my interest and imagination towards much of the subject matter covered in this thesis.

ACKNOWLEDGEMENTS

I would like to thank all of the helpful staff and classmates at the American Public University System. The wonderful people that I have been able to interact with during my graduate level studies have made me a much better student. I thank everyone that played a hand in making my education possible through the American Public University System.

Abstract:

The research examines computer generated predictive analytics and the possible future role of AI capacity quantum computers in that field of intelligence. Detractors of artificial intelligence (AI) have expressed their belief that computer technology is fundamentally limited to the role of a tool which human analysts may fruitfully rely upon while producing predictive analytics reports. Computer technology may be unable to ever exercise the problem solving skills needed to produce predictive analytics reports beyond mere pattern and trend recognition. However, machine learning and computer problem solving is advancing because of new methods of computation known as quantum computing. The emerging technology known as quantum computing is examined as an important factor that may allow computer technology to advance to the levels of AI needed so as to eventually produce complex computer generated predictive analytics reports.

TABLE OF CONTENTS:

Copyright, p………………………………………………………………………………….1

Dedication, p……………………………………………………………………………….2

Acknowledgements, p……………………………………………………………………..3

Abstract, p…………………………………………………………………………………4

Table of Contents, p………………………………………………………………………..5

Introduction, p…………………………………………………………………………..6-7

Literature Review, p…………………………………………………………………...7-38

Methodology and Research Strategy, p……………………………………………….38-41

Findings and Analysis, p……………………………………………………………...41-61

Conclusions, p…………………………………………………………………………61-66

References, p…………………………………………………………………………..67-69

Introduction:

The research topic considers if computer scientists may be able to invent quantum computers in the near future that will be capable of functioning at the levels of AI needed to independently produce predictive analytics reports which are of at least the same analytical quality as are now produced by human analysts. The research offers the following *research question*: in what ways can artificial intelligence based predictive analytics be advanced by quantum computing? This research question would be of interest to the intelligence community because of the possible advancements in the field of predictive analytics that would be achieved if quantum computing technology is reasonably expected to be someday capable of matching or exceeding human predictive analysts. Not all experts in the field of artificial intelligence are confident that computers will be developed with the ability to match human analysts; according to Chomsky and Danaylov (2013), the much respected linguist Noam Chomsky has stated his belief that computer technology will most likely not be able to solve problems in a way that is fundamentally the same as the problem solving cognition methods of human beings. Chomsky also expressed his belief that IBM's Watson is more of a "PR gimmick" rather than any kind of problem solving AI technology. Chomsky does not believe that the development of true AI is even remotely close to being accomplished in the foreseeable future (Chomsky and Danaylov 2013). However, the current inferior position of computer based predictive analytics in comparison to human based predictive analytics may someday be upset or reversed by the future development and application of quantum computing in the field of predictive analytics. The source by Cuthbertson (2015), explains that the processing speeds used in Big Data crunching would almost certainly be exponentially increased if quantum computing technology could be developed for commercial integration with a system such as IBM's Watson Analytics. Quantum computing uses quantum particles to express binary code, which means that the storage processing power would be much higher than any kind of non-quantum based computer technology. Limited testing has shown that quantum computing is possible and quantum computers also have the capacity to learn (Cuthbertson 2015). Indeed, despite the fact that quantum computing is still in a very early stage of development, promising signs of machine learning have been noticed by quantum computer scientists; quantum computing technology could eventually allow computers to perform at the levels of human analysts in the field of predictive analytics. The advancement of quantum computer technology could revolutionize the intelligence field of predictive analytics. Artificial intelligence is a subject of controversy, and some experts do not place a high degree of expectation that computer capabilities will advance enough to exercise unsupervised analysis near the abilities of human beings. While quantum computers may someday allow for such computer based analysis to take place, research is needed in order to help determine if such advancements are someday possible.

Review of the Literature:

The research concerns the unanswered question of in what ways quantum computers will benefit unsupervised computer based predictive analytics, and part of that larger question is examining the feasibility to actually design fully functioning quantum computers that exhibit a diversity of operational capabilities. Quantum computing is a newly emerging field of science. Uncertainties currently exist regarding the possibilities of creating quantum computers on a large scale. Challenges have been discovered by computer scientists that are studying and working on quantum computer technology. The question still remains if the emerging technology of quantum computers is indeed possible to create on a fully functioning level. This debate stems from the concerns over the instability of quantum particles. The issue is addressed by the source by

Fedichkin, Yanchenko and Valiev (2000), the issue of quantum de-coherence rates have been studied, and it would seem that qubits are indeed "coherent enough to perform error correction procedures" (Fedichkin, Yanchenko and Valiev 2000). Therefore, given that bright minds have been successfully working on quantum computer technology for years, and have made incremental gains in their endeavors, quantum computers will most likely be created in the near future. The resources mentioned above only reflect a very small fraction of the many works of research and other publications which readily exist that may be of benefit to the proposed research.

Quantum computer technology will most likely be advanced in the near future, but the limits of quantum computer technology is a serious issue among researchers. One of the primary problems in terms of quantum computer function is the unstable nature of sub-atomic particles, which can lead to errors often cropping up during quantum computer function. Computer scientists are working on ways to mitigate the problem of errors taking place when quantum computers are running; according to Hsu (2015), a primary challenge in the creation of a working quantum computer is the fact that the binary representing quantum particles known as qubits are notorious for inherent instability. Scientists are making progress in the endeavor to develop ways in which the quantum computers themselves will exhibit qubit error corrections. Google has funded a development team from University California, Santa Barbara in order to further advance quantum computer error correction; the researchers had already had success in terms of creating superconductor technology designed for quantum computing. The new project that Google is funding will allow for error correction of qubits to take place in a manner that would allow for a quantum computers to function without suffering from performance failures while running (Hsu 2015). Therefore, despite the fact that quantum computers are often prone to errors that end up jamming the performance of the devises, scientists have been successful thus far in early attempts at reducing such error problems. Quantum computer scientists could make more progress in terms of reducing quantum machine errors if more funding and resources were invested in quantum computer science.

Quantum computers use the entanglement process of sub-atomic particles in order to function. Quantum entanglement does not often stay in place when longer distances of communication are involved between quantum computer technologies; according to Zhong, Hedges, Ahlefeldt, Bartholomew, Beavan, Wittig, Longdell, and Sellars (2015), communication of around one hundred miles in distance via a theoretical quantum computer network is made difficult due to the fact that quantum entanglement tends to break down over such large distances. Designers of such a network could theoretically utilize a repeater protocol that preserves quantum information that may have been lost during the long distance transitions. The proposed quantum based repeater protocols would depend upon the concerned quantum entanglements having reasonably stable lifespans (Zhong, Hedges, Ahlefeldt, Bartholomew, Beavan, Wittig, Longdell, and Sellars 2015). Quantum computer networks may have the tendency to crash because of the inherent nature of quantum particles to change states. Classical computer technology may be more trustworthy in terms of running without incidents of crashes. Computer scientists are attempting to discover ways that would solve the problems that come along with developing quantum computers that can run without losing information every time the quantum particles inside of the chips change quantum states.

The manner in which quantum bits are encased inside quantum computer chips seem to have significant repercussions in terms of the stability of the quantum particles that are inside such chips; according to Zhong, Hedges, Ahlefeldt, Bartholomew, Beavan, Wittig, Longdell, and

Sellars (2015), researchers had been able to reach a maximum coherence time that lasts three hours by means of using a silicon-28 that works with a phosphorus donor. Zhong's team broke the coherence record and reached the six-hour mark by using a technique that utilizes europium-doped being placed inside yttrium orthosilicate during the creation and manipulation of the qubits. Quantum memory can be much more stable by using the europium technique pioneered by Zhong's research team because of the optical addressability offered by the matrix utilized during the quantum manipulations. Long range quantum computer transmissions tend to be best approached through the use of photons due to the fact that such optic based transmission is not difficult to create and send; but, ranges of communication that range from around a mile of more present challenges, such as light being absorbed, spread out, or absorbed during the lengthy path between the transmitter and receiver in a network (Zhong, Hedges, Ahlefeldt, Bartholomew, Beavan, Wittig, Longdell, and Sellars 2015). The more advanced studies that have been taking place concerning the medium by which quantum computers store and manipulate sub-atomic particles in entangled states has so far shown promise in regard to keeping the qubits in entangled states. Qubits existing longer in entangled states will allow for long distance communication via quantum computers; hence, cloud networks and Big Data could be theoretically stored and communicated through quantum computer technology in the future.

    The intelligence field of predictive analytics depends heavily upon the analysis of Big Data. Ying (2014), explains that the current era of Big Data presents challenges in the business world due to the chaotic volumes of information that business analysts must make sense out of in order to produce intelligence reports to consumers (Ying 2014). Therefore, with the rise of Big Data that seems to exist and increase during the current information age, the need for computer technology assisting human analysts is understandable. Because computers are naturally effective at processing the massive amounts of information present in Big Data, the trend in the intelligence field is to increasingly utilize computer technology for such tasks. The job of predictive analysts would be made easier if computers could offer predictions. However, differences exist in terms of how effective computer technology has performed in the varied aspects of the intelligence production cycle; computers seem to be particularly limited in terms of performing predictive analytics when compared to the performance abilities of human analysts.

Human beings are still much better than computers at predicting future events based upon the analysis of data; however, the analysis of Big Data will almost certainly entail humans utilizing computer technology in order to make sense of the ever increasing volumes of information that is being created and stored. The mass of information that exists inside Big Data may become increasingly overwhelming for human analysts to handle without some form of computer technology being utilized; according to *The Moore's Law of Big Data* (2013), the expansion of digital media that is part of Big Data is roughly keeping pace with the advancements in processor power predicted by Moore's Law. Moore's Law predicts that computer processing power will become twice as efficient within every two years because of the increasing advances in the construction and application of transistors inside a computer's processors. The increase of digitalized media that is included in Big Data may be increasing to twice as large roughly every two years, and such Big Data helps analysts discover truths about the world which is valuable to industrial research. The increase in processing power that has come with the fulfillment of Moore's Law over the years has given Big Data analysts better tools by which to find, record, and make sense of vast amounts of information (*The Moore's Law of Big Data* 2013). The exponentially increasing volumes of Big Data almost guarantees that on many occasions, human analysts will need to rely on computer technology in order to efficiently analyze Big Data, but uncertainty exists in regard to if computer processing power will continue to keep pace with the expansion of Big Data volumes. A possible future in which the volumes of Big Data exceeds the processing power of computers to effectively analyze may someday come into reality.

Processing power has thus far stayed true to Moore's Law. The continued advancement of computer processors may come to a grinding halt due to the realities of nanotechnology being used in computer circuits; according to Albert, Simmons, and Samoilov (2016), Moore's Law is a phenomenon regarding the units and processing speeds of CPUs, and has been a mainstay in the prediction that computer power will level up to twice as powerful roughly every other year. The continued validity of Moore's Law is coming into doubt; due to the fact that parts inside CPUs are becoming smaller and smaller, as the incredibly diverse amount of microscopic parts are becoming less likely to be made smaller and still function. Processor parts will eventually become smaller than atoms, and the laws of physics then switch from classical physics to that of quantum physics; hence, processors would have to eventually function in the realm of quantum physics as Moore's Law continues to progress (Albert, Simmons, and Samoilov 2016). The realities of the laws of physics may halt the continued and incremental increase in processing speeds that have occurred over the progression of computer technology; if processing speed stagnates, then the volumes of Big Data would theoretically become more difficult for computer technology to handle, and the field of predictive analytics would become less reliable in the future.

Computer engineers may be able to mitigate the problem of Moore's Law becoming an obsolete prediction if the nanotechnology used in processors could be made to work correctly at the sub-atomic level; according to the quantum computer scientist named Michelle Simmons (2012), the computer industry has spent very large funds in order to figure out how to decrease the size of functioning parts in processors. The components of modern CPUs are now nanotechnology, with parts in CPUs being many thousandths of times smaller than the thickness of a strand of hair. The incremental progression of miniaturizing computer technology has been a conscious effort; the computer industry has made sure to prove Moore's Law correct by scaling down the parts used in processors. The steady reduction in size of transistors inside processors has allowed computers to function with more and more processing power; modern processors that utilize billions of transistors working inside every chip. When researchers work on the predictable assumption that Moore's Law will continue to be proven right, as has been the case for decades, then by the year 2020, transistors will be as small as atoms. When transistors would eventually be miniaturized to the state of atomic size and then smaller, the realm of classical physics no longer apply, and the laws of quantum physics take over; due to the fact that sub-atomic particles function as waves as well as particles, the computational mechanics of classical CPUs would no longer function correctly. Moore's Law can be used to predict that without the advent of quantum computer technology, processing power can only advance for roughly another half-decade, and then stagnate. Moore's Law can only continue to be correct after another half-decade if quantum computers can be used to continue the advancement of processing power (Simmons 2012). Classical computer technology will almost certainly begin to stagnate in terms of processing power by the end of the decade; however, quantum technology may serve as a solution to the eventual problems of creating nanotechnology that works at the sub-atomic level. The continued validity of Moore's Law is essential if computer technology will be able to assist human analysts tackling Big Data; otherwise, practitioners working in the field of predictive analytics will increasingly encounter more and more difficulty attempting to produce predictive intelligence reports based on Big Data.

The challenge of sub-atomic nanotechnology development in computers could be a blessing in disguise if quantum computing is successfully advanced in the field of computer science; according to Simmons (2012), the plus side to the invention and use of quantum computers being used by society is that quantum based processors would theoretically be much faster than the classical processors used today. A classical processor's transistors can only perform one action at a time. Quantum processors utilize transistors that can perform different actions at the same time (Simmons 2012). Therefore, processing speed is much faster when a computer runs on qubits instead of classical bits. Quantum computers would theoretically allow for much faster processing power if the technology is able to be advanced. The logic behind why quantum computers would be much more powerful is that many computational steps could be completed in the same time intervals, and the fact that transistors would be able to perform more than one function at the same time, then the size of the chips could also be reduce; hence, allowing for faster processing and less space being needed for computer functions.

While modern classical computers run on a very large amount of transistors in order for computers to offer the processing speeds that exist today. Quantum computers would in theory, only need to run on a fraction of the transistors as classical computers need to run. The speeds of quantum computers would be adequate; according to Simmons (2012), every time a quantum bit transistor, known as a qubit, is added to a CPU, the processing power is made twice as powerful by just one added qubit, and this fact is expected to allow for exponential speedups in processing power as quantum computing advances in the future. Quantum computers are expected to become ideal forms of information technology that can be used to crunch Big Data, and thus be used by researchers to model economies and the environment; the vast amounts of unknown outcomes that are inherent in Big Data models would be effectively handled by quantum computers because of the inborn capabilities that qubit powered processors would possess (Simmons 2012). Quantum computing technology may be able to solve the problems inherent to the continued existence of Moore's Law. Quantum computers may prove to be much more effective than classical computer technology in a wide swath of applications. The possibilities of enhanced processing power that would be the result of quantum computers being further developed may lead to more advanced forms of artificial intelligence.

Computer technology has long been designed around the concept of binary language. Bits have served as the primary method that computers are able to process information in binary language; binary is a language that uses ones and zeroes in order to transmit and store information. The inherent superiority of quantum computing technology over that of classical computers lays in the difference between a bit and a qubit; according to Albert, Simmons, and Samoilov (2016), in the processors of classical computers, binary is expressed as either a one or a zero; a one is a small amount of electricity that is stored, and a zero is the absence of electrical storage. Binary is the hallmark method of processor functionality, and the segments of binary language that make up individual statements are called bits; binary language works at the macro-level of classical physics. Quantum-bits are abbreviated as qubits, and also express the language of ones and zeroes; however, the particles that are manipulated by a quantum computer in order to read and write qubits entail sub-atomic particles. The spins and polarized states of sub-atomic particles are used in the rendering of quantum binary computer language. Quantum binary methods allow for a computer to tackle a question by crunching parts of equations in parallel time; the use of qubits allow for computers to work out questions by performing computational measures all at once instead of part by part, and the speed of the processor is then much faster than a classical CPU. The parallel processes of a quantum chip is owed to the fact that a sub-atomic particle's spin can be positive, negative, and both positive and negative; a sub-atomic particle is not limited in terms of sequential reality that the macro-realm of classical computers must operate (Albert, Simmons, and Samoilov 2016). Quantum bits may be used to vastly increase processing speeds because of the laws of quantum physics allowing for faster computation. The application of quantum computing in the field of predictive analytics may allow analysts to make sense of Big Data in a way that classical computer technology cannot; however, because of the fact that quantum computing is still in the very early stages of development, the true potential of quantum computers being used in the practice of predictive analytics is not yet known.

Quantum bits can be designed by computer scientists, but the actual application of qubit processors into other areas of computer technology may prove to be a difficult task. Areas of uncertainty exist in the field of quantum computer development, and such uncertainty would have to be addressed before qubits could be more widely applied to computer assisted predictive analytics; according to Albert, Simmons, and Samoilov (2016), quantum bits can exist in super-positional states that allow for more processes in the chip to take place at once, and this is why a quantum computer would theoretically crunch data faster than a classical computer. Quantum bits do not need to operate independently of each other while crunching data in a CPU; instead, the laws of quantum mechanics allow for the qubits to be intertwined in states of entanglement, which is capable of further speeding-up a computer's processing capabilities. De-coherence is a major problem in terms of quantum computers; when super-positions break down due to interference from the outside world, a quantum computer loses the ability to properly function.

Great resources must be invested in order to meet the necessary environments that quantum computers need in order to correctly process information. The puzzle of how information in a qubit based CPU can be moved onto a hard-drive's ROM or RAM is another hurdle for developers; qubits are easily disturbed by the outside environment, and this means that moving information from the CPU of a quantum computer is a problematic step during such a computer's functions. Quantum computers are expected to someday be much more efficient than classical computers in terms of searching the World Wide Web for information and making sense of Big Data (Albert, Simmons, and Samoilov 2016). Qubits could theoretically revolutionize computer technology and allow for much more accurate predictive analytics reports, but while such advances may lay in the future, quantum computer technology has not yet been developed enough to fulfill such expectations at the present time. Quantum computer technology would be very useful for human analysts attempting to produce predictive reports based on the ever growing volumes of Big Data; however, faster processing speeds that may be available after the development of quantum computing would not necessarily be able to allow computers to exhibit the problem solving skills needed to produce predictive intelligence reports independent of human analysts.

  Human analysts are capable of producing predictive intelligence reports based on the analysis of Big Data. Computer technology has proved to be an effective tool for human analysts, but computer technology is certainly limited in terms of being able to produce predictive analytics reports without a human being involved in the analysis of Big Data; according to FICO's *How Does Predictive Analytics Differ from Data Mining and Business Intelligence?* (2006), there is a difference between the related disciplines of data mining and predictive intelligence. Human beings are needed to supply predictive intelligence, and although there are cyber-technology suites that are offered to customers that are marketed as being able to produce predictive analytic reports to consumers, such reports are merely based on the pattern recognition of commercial trends; computer generated predictive analytics reports are rather simplistic in nature (*How Does Predictive Analytics Differ from Data Mining and Business Intelligence?* 2006). This use of computers in the creation of predictive analytics reports is nonetheless being challenged due to recent computer technology advances. A report from IBM titled *What is Watson Analytics?* (2015), claims that the cloud based service known as Watson Analytics offers its customers a low effort and computer generated means of obtaining useful predictive analytics reports which enables users to make wise decisions for the futures of their endeavors (*What is Watson Analytics?* 2015). The nature of Watson Analytics is also a subject that has been covered by previous research. Watson may be merely able to retrieve data in order to answer questions faster than human contestants in a trivia gameshow. Rapid data retrieval and information linkage is not necessarily the same function as AI based problem solving skills.

The questions over whether or not Watson analytics is actually a form of artificial intelligence should be addressed. Artificial intelligence is understood by experts to entail being able to think in a way that is similar to a human being. Computer scientists that have been active in developing Watson made efforts to ensure that Watson's processing of information was more sophisticated that the simple retrieval of information that earlier predictive analytical technology relies upon in order to serve consumers; according to Lohr (2011), the US Department of Defense funded the creation of Watson Analytics. Watson Analytics is inspired by the neuronal functions found in the human brain, and this specific neuronal design is believed to be a primary reason why Watson is capable of analytical skills (Lohr 2011). The evidence clearly indicates that IBM's Watson displays functions which are beyond mere data retrieval based on information input. The possible AI functionality of Watson may be due to a computation design loosely patterned after the human brain; according to Siegelmann and Sontag (1991), the concept of creating computer technology that is designed as a reverse engineered brain that uses binary logic gates which function in a manner similar to neurons has been researched for decades (Siegelmann and Sontag 1991). Watson may indeed possess true problem solving capabilities which are needed to produce unsupervised predictive analytics reports. Watson does possess dynamic predictive functionality; however, whether or not Watson is capable of AI capacities is still a matter of debate. The problem solving capabilities which are inherent in Watson's system is almost certainly far less advanced than a human beings problem solving abilities.

The simple fact that IBM's Watson has been created with a computation design inspired by the human brain does not necessarily mean that the actual functionality of Watson fundamentally resembles the cognitive functions of a human brain. Watson's problem solving capacities is still a matter of debate among experts in the field of AI; according to Chomsky and Danaylov (2013), Chomsky argues that the human brain works intrinsically unlike any of the current attempts at AI. Computers will most likely not be able to solve problems on their own during the current age of computer development. Chomsky grudgingly admits that some computer technology may be capable of exhibiting some problem solving skills, but he stresses that computers are limited and primitive in respect to true problem solving abilities of a human brain (Chomsky and Danaylov 2013). The debate over Watson's true capacities and usefulness is ongoing, and certainly IBM's Watson has been criticized as being little more than a useful prop used by human analysts. The ability to practice problem solving skills is not as advanced in computer technology such a Watson Analytics as is the human brain's abilities to do so. Critics of IBM's Watson may be unfair in terms of belittling that technology's usefulness to analysts.

Advocates of Watson's possible AI capacities may have over emphasized the product's AI characteristics as a result of market realities; customers of predictive technology that have been told that a product is capable of exhibiting AI level functionality may be more inclined to buy that product.

Computers capable of exercising artificial intelligence may be invented in order to exercise problem solving abilities, the application of Bayesian networks into such computer technology may be one of the essential elements needed for that cyber-advancement. The source by Charniak (1991), explains that researchers and theorists in the fields of probability and uncertainty have increasingly been embracing the methodology known as Bayesian networks as a means to eventually design computer technology to function as AI, and this would theoretically enable such computer technology to predict outcomes based on incomplete information (Charniak 1991). Bayesian networks methods of analysis can be a complex discipline, which is known to use what are known by theorists as click trees, which may be very useful in the emerging field of AI. The source by Mengshoel (2010), explains that Bayesian networks are enabled by the use of methods called clique tree clustering as well as propagation; which entails propagation that takes place by way of clique tree propagation that is composed out of a Bayesian network (Mengshoel 2010). The methods by which Bayesian networks reach conclusions have been established for some time. Probability theorists have argued that such methods could be useful in the advancement of computer technology that may someday be able to produce predictive analytics reports on at least the same quality level as human analysts. Bayesian networks may someday be incorporated to help quantum computer technology make AI level analysis.

The use of Bayesian networks being more heavily integrated into computer technology such as Watson alone may not allow computers to advance to the AI levels needed to independently practice predictive analytics. The processing of Big Data may be too much of a task to allow Bayesian networks to be fully and properly worked out by computers. When a very large amount of information needs to be worked out in a Bayesian formula, there simply may not be enough time for analysts to wait for computer outputs. The development of quantum computer technology may give a solution to the problem of inordinate time that is sometimes spent analyzing Big Data; the source by Cuthbertson (2015), explains that the emerging technology of quantum computing will be able to someday dramatically increase computer speeds and advance machine learning (Cuthbertson 2015). Therefore, when considering the value of Big Data being analyzed by way of Bayesian networks, the advantages of using quantum computers for such tasks is obvious; the faster the processing time a quantum computer is able to perform, the more Big Data can be analyzed through methods such as Bayesian networks; artificial intelligence may be approached through such combined methods, and consumers could greatly benefit from improvements in predictive analytics reports.

Machine learning is made possible by algorithms which allow for computers to adapt or learn. Computers are often not dependent on constant human supervision and re-programing; according to *Machine Learning* (No date), machine learning is when a computer is capable of discovering insights about the world based on analyzing data that is stored inside the computer's memory. Computers which lack machine learning capabilities cannot evolve to changing circumstances and realities that pertain to the data which the computer receives and analyzes; such computers have no ability to amend conclusions and will depend on human beings to fulfill that role. Machine learning is an established concept that has gained popularity in recent years because of the developments of more advanced algorithms that display the ability to process and analyze Big Data much faster and more efficiently than older machine learning algorithms. Recent forms of technology that is being marketed today; self-driving cars that are designed by Google, algorithms which analyze tracking cookies in order to more efficiently market advertisements to potential customers, and algorithms that are designed to analyze discussions that take place over social media, or detect fraud that is taking place are all examples of machine learning. The increased speed and processing power associated with modern machine learning algorithms is enhanced by the application of linguistic rule creation, the computer's capacity to rapidly process streams of data in order to create working data models, and Bayesian analysis (*Machine Learning* No date). Therefore, machine learning is not achieved by the simple application of one form of technology or scientific discipline; a synergy of different scientific techniques must be applied in order for machine learning to be achieved. The many ways in which machine learning could benefit society is impressive, but further advances appear to be needed in order for such applications of machine learning to become wide-spread.

    The application of machine learning in the field of predictive analytics seems very plausible. Big Data analysis is a fundamental component to predictive analytics. Computer algorithms that are written for the purposes of machine learning may be applicable towards the tasks that entail crunching Big Data; according to *Machine Learning* (No date), the growing amounts of Big Data have increasingly allowed machine learning algorithms to analyze more data, and hence give more informed outputs to consumers. The writing of advanced machine learning algorithms that are designed to work in conjunction with Bayesian analysis methods and linguistic rules have brought about computer based predictive analytics that give outputs to customers; while human beings are needed in the design and development of such machines, humans are not needed during the computer's analysis and prediction output functions (*Machine Learning* No date). The understanding among experts in the field of machine learning is that the faster new technology is able to achieve information processing, the more advanced and dependable a machine running a learning algorithm can be for consumers. The advantage to using new quantum computer technology in the near future is the increased processing speed which could be offered during machine learning.

Machine learning algorithms seem to allow computers to learn from new data in a manner that is not dissimilar to how biological brains learn from the external environment; according to Domingos (No date), computer scientists have created special algorithms which amend the behavior of computer programs based on heuristic based modification, which is essentially known as machine learning; computer algorithms can be written that are designed to expect that patterns and trends will play out in similar manners as have past patterns and trends; and hence, these machine learning algorithms are able to give predictive advise to human consumers. Machine learning algorithms have been widely used in the marketing of commercial products, the designing of medicine, and cyber-security (Domingos No date). While machine learning algorithms may benefit a host of technology and intelligence related fields, such algorithms do not appear to possess the ability to come up with original conclusions; everything that a computer can learn while running on a learning algorithm is strictly related to past trend patterns. Human analysts would seem to possess a comparative advantage in terms of innovative thought over computers that run on learning algorithms.

The fact that a machine learning algorithm is able to practice a reasonable amount of autonomy may be encouraging for advocates of the possibilities of AI. The use of machine learning algorithms may be of benefit to human analysts that can rely upon computer technology to handle much of the bulk of Big Data; according to Domingos (No date), machine learning algorithms are able to change and adapt in accordance to changing data patterns which are analyzed by those algorithms, and human beings are not needed in order for machine learning algorithms to evolve in response to new data patterns. Algorithms for machine learning are heavily utilized in the fields of data mining and predictive analytics. Algorithms are widely designed to operate as classifiers; a classifier essentially takes in data for analysis, and upon the completion of the analysis of the data flow, gives an output that is used by the human consumer in order to make informed decisions; the classification process is often times accomplished by means of Boolean vector equation templates that can be modified to analyze different data streams which are fed into a computer for subsequent analysis by the machine learning algorithm. Computer algorithms can be designed to heuristically make generalizations about new data streams, and such generalized expectations use pattern recognition that the learning algorithms find in older data streams. The learning algorithms are able to learn from past data stream patterns and trends in order to make predictions about trends and patterns found to be associated with new data streams (Domingos No date). Therefore, developers of machine learning algorithms would be able to make significant advances in the field of predictive analytics if the speed and flexibility of computer processing could be enhanced, since quantum computers may offer such advantages, there appears the very real possibility that machine learning would reach much more dependable and complex levels of performance once quantum computer technology is widely introduced in the field of machine learning.

   Machine learning algorithms may at times have trouble correctly analyzing and adapting to vast amounts of diversified Big Data. Quantum computers are expected to thrive and reach better performance levels when running and analyzing such vast data streams through a learning algorithm; according to Lloyd, Mohseni, and Rebentrost (2013), the classification of data stream vectors by use of machine learning algorithms tend to rely upon operations of equations that concern the numerical and special dimensions of vectors. Quantum computational methods are particularly effective at manipulating vectors that are composed of complex dimensional characteristics, researchers believe that future machine learning algorithms which would incorporate quantum computing techniques would exhibit higher learning capabilities than the current non-quantum machine learning technology used today. Machine learning algorithms typically perform best when such algorithms are given Big Data in the form of tensor products that are composed of vectors and vectors of data that are fed as a data-stream (Lloyd, Mohseni, and Rebentrost 2013). Quantum computing and learning algorithms are potentially very compatible with each other; both forms of computer technology are beneficial in terms of crunching large amounts of information. The future of predictive analytics may someday be altered for the better by the coupling of learning algorithms and quantum computation; however, difficulties may arise when computer engineers attempt to figure out methods of designing qubit processors to actually be able to run complex learning algorithms, which may or may not be feasibly possible.

The advantages to eventually switching over to quantum computers in order to crunch data is somewhat obvious, and the field of predictive analytics would most likely benefit from the incorporation of quantum computer technology; according to Lloyd, Mohseni, and Rebentrost (2013), quantum computational methods are inherently better at analyzing Big Data than traditional computational methods; researchers believe that quantum based machine learning algorithmic methods will be more effective at learning than current traditional machine learning algorithms that do not incorporate quantum computing. Quantum computers would be able to store data in the q-RAM so as to be read in the form of sub-atomic particles, which is then run through a quantum based learning algorithm; at which point, the quantum computer's machine learning algorithm would be notably faster than what would normally ever be performed by a traditional computer's machine learning algorithm processing time (Lloyd, Mohseni, and Rebentrost 2013). Therefore, when considering how quantum computers would work in support of learning algorithms, the outlook appears very positive in terms of the beneficial role that quantum computers would play in the field of predictive analytics, and this is because of the strengths that would come along with using quantum computers to analyze very large amounts of information. While classical computer technology is capable of crunching Big Data, quantum computers would probably be better for the task; the difficulties appear to lay in actually developing the needed quantum technology that could be used to analyze Big Data.

The architecture of classical computer technology, such as RAM and algorithms written in coding language appear to be the primary designs in which quantum computer scientists have been planning to devise quantum machine learning; however, the architecture of the human brain may be an optimal model to examine for quantum computer scientists. Researchers have speculated that the human brain is in fact a biologically based quantum computer, and that designing quantum computers in a way that resembles human brain structure could exponentially improve artificial intelligence performance; according to Rigatos and Tzafestas (2006), if human brains use some form of biologically based quantum computational functions, such as brain synapses playing the role of fuzzy variables during brain processes, then there is the very real possibility that the human brain runs quantum based learning algorithms wherein associative memories may exist in super-positional states. Human brains may utilize quantum super-positional states to store and associate memories together in meaningful patterns by way of forming vectors of data that are represented in the form of sub-atomic particles which form relational attractor patterns by means of unitary rotations (Rigatos and Tzafestas 2006). The fact that the human brain is in overall terms, much more effective at such tasks as learning and problem solving may be a result of the dynamic methods by which the brain stores memory; instead of using a static method such as computer RAM, the brain may in indeed be utilizing quantum particles to store and learn associations between memories.

Theories which speculate that the brain utilizes quantum particles to store and make sense of memories is new, and many questions still remain to be answered in regard to the validity of such claims, as well as the methods by which the brain may be accomplishing such a task; however, if such a cognitive phenomenon is being performed by the human brain, then the superiority of human intelligence over computer technology seems easier to understand; according to Rigatos and Tzafestas (2006), traditional computers which are non-neuronal inspired are inferior to the human brain in terms of such tasks as storing and associating memories. Scientists speculate that if new computer technology were created which essentially take the form of synthetic neuronal design and functionality, then such new computers would almost certainly be much better at associating and constructing meaningful patterns of memorized data (Rigatos and Tzafestas 2006). Quantum computer engineers may need only observe and understand the quantum level functions of the brain in order to design artificial intelligence. The theory of brain science which has proposed the existence of the brain's quantum level cognitive functions still needs to be proven, but the endeavor to more fully understand the brain's quantum behaviors could be a positive learning experience for brain scientists and computer scientists alike.

Quantum computers would benefit in terms of functionality by being modeled after quantum brain functions, and in order for quantum computer scientists to achieve such an accomplishment. The neuronal architecture of the brain's biological networks would most likely have to be synthetically replicated by quantum computer engineers; according to Rigatos and Tzafestas (2006), neural inspired quantum computers would be better at learning as a result of mimicking the enhanced memory associative capabilities that are believed to be fundamentally exhibited by the human brain (Rigatos and Tzafestas 2006). The debate over whether or not computers could ever be designed to function at the level of AI may end up being settled if computers are someday designed to function in a capacity that resembles the human brain, as well as any quantum cognitive functions which take place in the human brain; therefore, if the speculations are indeed true that human brains utilize quantum particles in order to process information about the external environment, then the next obvious step in terms of the advancement of machine learning would be to further develop quantum computers that are inspired by the human brain.

The human brain is increasingly becoming a source of guidance for computer scientists; developers hope to create more advanced computer technology by learning from brain functions during cognitive processes; according to Angelica and Schmidhuber (2012) the Swiss Artificial Intelligence Lab (SAIL) has designed computer technology that is designed to exhibit machine learning capabilities. The SAIL has led pioneering efforts to design computer technology that is inspired by the neuronal functions of biological brains. SAIL has tested the neuronal inspired computers in imagery pattern recognition tests as a way to ascertain if machine learning is being advanced by way of designing computer technology in a manner that is similar in some respects to human brains (Angelica and Schmidhuber 2012). The incorporation of neuronal cognition designs into computer technology has proved to be an effective pioneering effort on the part of computer developers. The computer technology that has been invented by the Swiss Artificial Intelligence Lab has been classical computing instead of quantum computing; however, a synergy between neuronal inspired computers and quantum computers may allow for computer science to advance to new heights of performance capabilities.

Machine learning and problem solving capabilities can be compared to that of the human capacity to learn. The Swiss Artificial Intelligence Lab's computer technology has been tested against human beings; according to Markoff (2012), machine learning capabilities can be tested by comparing the ability of experimental computers to learn against that of human contestants. Recent tests indicate that humans and computers are about equal when learning image patterns are involved in specific tests. Machine learning has been advanced by designing new computers with processing functions that resemble that of the human brain (Markoff 2012). Technology such as the computers designed by SAIL have proven that computers can exhibit some basic machine learning and problem solving capabilities; however, such abilities have only been shown by computer technology in controlled situations such as tests. The random nature of Big Data may present serious challenges to advanced computers, even if a computer product has performed adequately in controlled test situations. The task of analyzing Big Data may entail a high degree of random and unexpected variables in many instances, and such situations may confuse advanced neuronal inspired technology. Quantum computer technology may allow faster computer processing; however, whether or not quantum speed-up would give neuronal inspired computers AI capabilities is still unknown. Neuronal inspired computer designs and quantum computing would most likely be of significant benefit to the field of predictive analytics. The exact synergy between neuronal inspired computing and quantum computing is yet to be seen.

IBM's Watson is one of the most prominent examples of possible AI based predictive analytics products. There is the marketing initiative on the part of IBM to claim to the public that their corporation has invented computer technology which is capable of generating predictive analytics reports to customers without depending upon human beings beyond the products design stage and data collection. Whether or not this claim is true may be determined by an examination of literature that covers the creation and performance functionality of IBM's Watson. The source from Thompson (2010), explains that IBM's Watson computer system proved itself to be better at winning the game show "Jeopardy!" than its human opponents; the functionality of Watson is in some ways similar to the functions of a biological brain (Thompson 2010). Therefore, Watson may indeed be performing some basic problem solving abilities needed to perform predictive analytics; however, some critics may justifiably question whether or not winning a game of "Jeopardy!" actually necessitates problem solving skills or not. There remains the possibility that Watson is nothing more than a computational prop designed to display to onlookers the appearance of AI. Quantum computer technology is not offered by the current version of Watson, but there is no reason found in the literature to believe that quantum computational functions would not be of great benefit to future versions of that analytical product offered by IBM.

The Turing Test is a popular method of ascertaining if computer technology has yet been invented that could be categorized as artificial intelligence. The Turing Test can be carried out at different times and places, and the execution of the testing methods may be in the hands of the event promoters and managers of such testing events; according to a report from the University of Reading titled, *Turing Test Success Marks Milestone in Computing History* (2014), the Turing Test 2014, was an event in order to ascertain if artificial intelligence has yet been created by computer scientists in which computer programs are tested to be able to carry on conversations with people in order to be as convincing in conversation as a normal human being. The judges at the Turing Test 2014 were to determine if the conversationalists hosted at the event were humans or computers (*Turing Test Success Marks Milestone in Computing History* 2014). The ability for a computer to carry out a lucid conversation in real time with a human judge would require some fundamental elements of artificial intelligence; words that are used in language can often times possess different meanings based on the nature of a conversation, and this reality can make the task of designing a conversationalist computer very complex. The test designers may place some reasonable boundaries on what was allowed by the judges to be said in the conversations, or the tests could be limited to whatever the judge wants to ask the computer or person on the other side of the conversation; however, the Turing Test is meant to place practical standards of conversational abilities upon the computers that are entered in such events.

Human beings may be more gullible than some others in the course of a conversation, and one judge at a Turing Test event may be much more observant and skeptical than a different judge that is also communicating with both humans and computers. The use of multiple judges in Turing Test events makes good sense; according to *Turing Test Success Marks Milestone in Computing History* (2014), a Russian team designed a winning computer program which they named "Eugene" that mimicked a teenager when competing in the Turing Test 2014; the test's criteria for passing depended on if any of the computer programs were able to trick the human judges thirty percent of the time into believing the two way conversation consisted of two human beings instead of a human and a computer program. Teams that have created computer programs that competed in the Turing Test have not all been winners, but the Russian designed "Eugene" became a winner because the program was able to trick thirty-three percent of the judges that they were communicating with a human being instead of a computer program (*Turing Test Success Marks Milestone in Computing History* 2014). The Eugene computer program may not have been a able to perfectly mimic the ability of the standard human being in conversation, but the computer program was able to seem human enough to trick a sizable minority of human judges; the likelihood of more advanced computer programs being built upon the Eugene design is reasonably certain, and such new programs would probably possess performance improvements over Eugene.

The computer programs that compete in Turing Test events may be incrementally more and more advanced as the years go forward; this would make sense given the fact that computer technology has been advancing over the past several decades. The advancing models of computers that are being submitted for competition in Turing Test events may give the event coordinators the chance to curtail the events in a way that allow for greater validity results when determining if a computer is capable of artificial intelligence; according to *Turing Test Success Marks Milestone in Computing History* (2014), the Turing Test 2014 was conducted in a manner that was much more difficult for computer programs to pass; the conversations did not have restrictions imposed in terms of what could be asked to the programs, which raised the standards in terms of difficulty level for the computer scientists trying to create passable programs; such standards were in line with computer visionary Alan Turing's original testing standards, and the program "Eugene" was the first computer product that ever passed a genuine Turing Test in recorded history, and is thus considered under the parameters of the Turing Test to be artificial intelligence (*Turing Test Success Marks Milestone in Computing History* 2014). The Turing Test 2014 winner called Eugene and IBM's Watson both appear to have been designed in a way that conforms and responds to human spoken language vernaculars; language acquisition may be a corner-stone of artificial intelligence attainment, which could explain the impressive performances of both Watson and the Eugene program. Quantum speed-ups during the processing of Big Data by way of learning algorithms would almost certainly bring about significant advances in future programs that will be entered for competition in the Turing Test.

Quantum computing is still a very new field, and there does not exist a particularly large amount of functioning quantum computer products that exist during the present time. D-Wave is possibly the most prominent example of a quantum computer product that currently exists in the world today; according to Metz (2015), the D-Wave is different from the typical theoretical designs of quantum computers. D-Wave is typically only useful in terms of solving optimization problems that involve equational tunneling. D-Wave utilizes quantum annealing in order to solve problems. The computer design known as D-Wave is unique in the sense that the quantum bits actually run on super-conductor technology. Constant energy flow in a circular pattern is used in D-Wave's quantum bit design (Metz 2015). The D-Wave design is certainly innovative in terms of how the machine utilizes quantum particles in order to run qubits; however, D-Wave's reliance upon quantum annealing may not have been the best approach for the machine's designers to have taken. The fact that D-Wave is limited to the solving of optimization problems may be an indication that quantum annealing may not be the best method to design qubit based processors.

The D-Wave's capacity to solve optimization problems may be of assistance to predictive analysts that process Big Data. The D-Wave's specific optimization problem solving capabilities may be of great use when Big Data needs to be crunched in order to produce predictive analytics reports; according to Brandom (2014), the D-Wave has been commercially advertised as a quantum computer that can crunch Big Data via a phenomenon known as quantum speed-up.

The debate as to whether D-Wave is a true quantum computer has been put to rest; D-Wave is quantum because of the fact that quantum annealing is used during D-Wave's processes. The debate as to whether or not D-Wave is faster than classical computer technology is still ongoing (Brandom 2014). The D-Wave's use of quantum annealing in the operation of the machine's qubits shows that D-Wave is a true quantum computer. The D-Wave can only be truly valuable to Big Data analysts if that computer is truly able to exhibit quantum speed-up; otherwise, D-Wave may only be at the same usefulness to analysts as classical computer technology. The cost of developing, buying, or maintaining a quantum computer such as D-Wave is another factor that buyers would possibly want to consider before investing in such a quantum machine. Marginal amounts of quantum speed-up over the performance of the best classical computer technology is exciting from a purely scientific point of view, but in order for a quantum computer to be further developed, the market for such a machine would logically also have to exist. The quantum speed-up on the D-Wave may have to be significantly noticeable in order for the developers of that computer to receive adequate funding needed for further development.

One of the main problems that occur when computer scientists attempt to detect the phenomenon known as quantum speed-up is being able to understand what exactly took place during a quantum computer's run time during performance tests. Sometimes understanding what exact speed a quantum machine such as the D-Wave is running at during tests is complicated; according to Ronnow, Wang, Job, Boixo, Isakov, Wecker, Martinis, Lidar, and Troyer (2014), the D-Wave has been tested against classical computer technology in regard to processing speed. The researchers that conducted the tests between the D-Wave II and classical computer technology have speculated as to whether or not that version of D-Wave exhibited quantum speed-up during those particular tests. The debate over D-Wave's quantum speed-up or lack thereof, is in regard to acute periods of computer performance, and not in concern to the general speed of D-Wave's processing power in comparison to the classical computer technology that was pitted against that quantum machine (Ronnow, Wang, Job, Boixo, Isakov, Wecker, Martinis, Lidar, and Troyer 2014). The best way to test D-Wave is to compare that machine to classical computer technology; however, the exact tests and methods of interpreting test results may be complicated. Validity concerns over whether or not quantum speed-up is being exhibited by D-Wave, and if the tests are measuring what needs to be measured may be issues of debate among researchers involved in D-Wave performance tests. The D-wave's quantum speed-up capabilities are thus far shrouded in debate and confusion.

D-Wave has been produced in different versions, and the new version of D-Wave has also been tests in a performance test against classical computer technology, and the expectations that the new version of D-Wave can exhibit enhanced performance over the older versions has been high. The D-Wave has been invented to be able to calculate answers to specific types of problems, which involve equations involving tunneling. The D-Wave has been designed to run on special algorithms during run times, and those special algorithms are intended to allow for the machine to achieve quantum speed-up; according to Denchev, Sergio, Boixo, Isakov, Ding, Babbush, Amelyanskiy, Martinis, and Neven (2016), optimization problems that a quantum computer such as D-Wave may solve typically involve calculating how to best tunnel through walls of energy. Classical computer technology can run algorithms in order to simulate calculations for certain types of tunneling problems. The newer quantum machine model known as D-Wave Two-X has been compared in efficiency and speed tests against classical computer technology; the question that the researchers wanted to discover in the tests was if D-Wave's quantum annealing processing methods are capable of exhibiting quantum speed-up and out-performing classical computer technology (Denchev, Sergio, Boixo, Isakov, Ding, Babbush, Amelyanskiy, Martinis, and Neven 2016). One prominent question among the computer science community is how much speed-up will be offered by the development and application of quantum computers in the effort at crunching data. The method of quantum based computation that takes advantage of the phenomenon known as quantum annealing may prove to be an efficient design for achieving quantum speed-up; otherwise, D-Wave may someday lose popularity as a quantum machine model. Brilliant individuals will most likely continue to work towards advancing quantum computing even if D-Wave is no longer being developed.

Microsoft has interested in the subject of artificial intelligence, and wants to know if the D-Wave would be of use in further developing AI technology; according to Brandom (2014), Google has bought a D-Wave quantum computer that uses quantum annealing during the system computational functions; Google worked with Microsoft to test the performance of their D-Wave computer against some of the most advanced non-quantum computers that currently exist on the market today. Google is currently continuing its development of the D-Wave II system, which is hoped by the researchers working at Google's Quantum AI Lab to exercise significant quantum speed-up when officially tested. The use of quantum annealing as a method as a method is uncertain to yield notable quantum speed-ups (Brandom 2014). The fact that a pioneering quantum computer such as D-Wave, that uses the controversial method of quantum annealing as the primary method of computation shows that even a clumsy prototype of a quantum computer is as noteworthy as the very best traditional computer technology currently on the market. The test run by Google and Microsoft concerning the D-Wave may suggest to optimists that advancements and improvements in the field of quantum computer technology would most likely lead to the phasing out of traditional computers on the market place.

D-Wave's use of quantum annealing may not be the best method in terms of quantum manipulation for computational purposes, and D-Wave is yet to be proven as inherently superior in comparison to traditional computer technology. A quantum computer research and development team has achieved a limited but real example of machine learning through the use of quantum computational methods; according to *First Demonstration of Artificial Intelligence on a Quantum Computer* (2014), quantum computers can manipulate an atomic nucleus to solve calculations. Quantum bits can even solve different problems in parallel time. Every time a qubit is added to the processing power of a computer, the power of that computer is made twice as strong (*First Demonstration of Artificial Intelligence on a Quantum Computer* 2014). Quantum speed-up may be an essential element that computers would need in order to exhibit AI functionality. A quantum computer has been tested in order to ascertain if that particular machine is able to perform machine learning. The source by Li, Liu, Xu, and Du (2015), explains that computers using quantum parallelism based algorithms would theoretically be capable of functioning in the capacity of artificial intelligence. Quantum technology using four qubits has demonstrated the ability during pioneering testing, to learn how to recognize hand-written characters (Li, Liu, Xu, and Du 2014). Therefore, the role of quantum computer technology may be an essential element needed in terms of designing computers that can produce predictive analytics which draw conclusions based on machine learning and problem solving methods; machine learning is a mandatory characteristic that computers need in order to perform artificial intelligence driven predictive analytics, and quantum computers may be better at machine learning than traditional computers.

The currently existing literature that has been reviewed by the research is abundant, but there nonetheless exist many gaps in knowledge at the present time. The major issue in question is if quantum computers will eventually be invented in a way that exponentially outperforms traditional computer technology. Quantum computer technology certainly appears to be possible, and already has been shown to exist in limited forms. Classical computer technology is possibly able to perform some basic artificial intelligence level skills; however, the question of how much assistance quantum computer technology will be to the effort to invent higher level artificially intelligent computers is not fully understood by the literature at this time. The analysis undertaken in this research intends to further answer the question of how much further quantum computing will advance artificial intelligence that can be used in the field of predictive analytics. The gaps in knowledge that happens to be present in the existing literature on the subject leads to the following *Research question*: in what ways can artificial intelligence based predictive analytics be advanced by quantum computing?

A major question that the literature is able to address but fails to answer is if quantum computer technology can be invented in the near future that is inspired by a biological brain's neuronal design. Questions currently exist as to whether or not the human brain uses any form of quantum particle phenomenon in order to function; if the answer is found to be affirmative, the obvious option for computer scientists would be to model quantum computers after the brain's quantum methods. The research intends to fill in the gaps in knowledge that exist in the literature; while completely finding the answers to such gaps in knowledge are not currently possible in the present time, the research does hope to help further answer the above mentioned uncertainties.

Methodology and Research Strategy:

Despite the large amount of literature that currently exists which the research is able to draw upon, the primary gap in knowledge that needs to be filled is expressed in the research question: in what ways can artificial intelligence based predictive analytics be advanced by quantum computing? Despite the fact that some experts in the fields of predictive analytics, AI, and quantum computing have hopes that quantum computing will someday enable computer technology to produce predictive analytics reports at the same performance levels as is found among human analysts, no one truly knows if quantum computing will be capable of advancing computer technology to that particular capacity. In order to answer the research question being pondered in this research, the research's associated hypothesis will be tested by the operationalization of independent and dependent variables. The impact of the independent variable upon the dependent variables will be measured in order to prove or disprove the proposed research's hypothesis. The <u>independent variable</u> used in the research is that of quantum computer technology that is being applied or is believed possible to be applied to AI designs that could be used in the field of predictive analytics. The <u>dependent variables</u> used in the research are computer based problem solving capabilities, probability outputs, and machine learning capabilities that can be applied to the field of predictive analytics. Upon the event that the research finds that the independent variable significantly impacts the dependent variables examined in the research, then the hypothesis will be found as supported. However, if the impact of the independent variable upon the dependent variables is found to be insignificant/negligible, then the hypothesis used in the research will be found to be unsupported.

   The variables will be operationalized through the examination and analysis of selected case studies. The research's control group case studies will concern advanced classical computer technology. The test group case studies will concern the problem solving skills, probability outputs, and machine learning capabilities of quantum computer technology that can be used in the field of predictive analytics.

The impact of the independent variable upon the dependent variables will be measured in the research by the examination and subsequent analysis of any performance enhancement that is evident with the quantum computer technology when compared to the non-quantum based computer technology examined in the control group case studies. The test group will entail selected case studies concerning quantum computer technology being utilized in computer performance that could be used in the field of predictive analytics. The control group will entail the examination of case studies that concern computer technology performance that is non-quantum based that can be used in the field of predictive analytics. Qualitative and quantitative measurements may be applied in the research's operationalization of the variables.

The case studies used in the operationalization of the variables will be collected for the purposes of the research from such sources as academic journals, articles, interviews, reports, as well as other reliable and trusted sources of information that fall into the parameters of the research's case studies. Every case study examined will be selected on the grounds of being compatible and applicable to the operationalization of the variables.

The case studies will be analyzed in order to operationalize the variables, and thus test the research's hypothesis. The hypothesis has been devised in a manner that will allow for proper testing to take place, and through that testing, the hypothesis will be supported or found to be unsupported. The research places forth the following hypothesis: despite computer based predictive analytics technology currently being only capable of producing reports that are of a quality which lack the analytical problem solving capacities of human analysts, the development and subsequent applications of quantum computing technology will bring computer driven predictive analytics to at least the same level of analytical problem solving capabilities as is routinely offered by human analysts.

When the hypothesis has been properly tested, the most advanced computer technology that is being used or is emerging but is most probably applicable to the field of computer based predictive analytics will be considered in light of the tested hypothesis. The results found after testing the hypothesis and the consideration of how those results will have relevance upon the most advanced computer technology that is being developed that can be used in predictive analytics, will then lead to answering the <u>research question</u>: in what ways can artificial intelligence based predictive analytics be advanced by quantum computing? The testing of the research's stated hypothesis will allow for answers to be reached in regard to the research question.

Certain limitations and biases are expected to accompany the case studies themselves. Validity problems that may limit the proposed research may be in the form of market realities that encourage the positive acclamation of the quantum computer performances on the part of the authors and reporters (so as to increase future funding); test failures are less likely to be published or reported upon when compared to successes in emerging computer technology tests. The study will face limitations in regard to the fact that emerging computer technology such as quantum computing and very recent developments in other computer based predictive analytics technology is still only in testing phases of development; therefore, the true potential and limitations of the technology that will be examined and analyzed in the case studies cannot be fully known at the time of the operationalization of the research's variables.

Findings and Analysis:

In order to operationalize the variables for the purpose of testing the hypothesis, the research turns to the examination and analysis of the selected case studies. The first case study concerns comparison tests between a quantum computer design known as the D-Wave and classical computer technology. The need for a separate section examining the control group case study does not exist in regard to case study one; this is because of the fact that the research is able to treat the performance results of the quantum computer design examined in the first case study as the test group, and the performance results of the classical computer technology that was tested against the D-Wave serves the purpose of the de facto control group. The second case study entails the test results of a different quantum computer design that happens to function on only four quantum bits; that particular quantum computer design is taken into consideration alongside a control group case study that involves the performance results of a revolutionary classical computer design that was tested to perform similar tasks as the four quantum bit computer has been designed to perform.

Case Study One:

The D-Wave quantum computer will be examined in terms of any potential quantum speed-up that the machine may be able to exhibit. The question of whether or not a quantum computer can out-perform classical computer technology is very useful in terms of operationalizing the variables used in the research. In order to appreciate the test results, a general understanding of how D-Wave operates is necessary. Classical computer technology is not limited to only one design method; likewise, quantum computers can be invented with a diversity of designs. The logical assumption is that some quantum computer designs will be more effective than others; according to Metz (2015), D-Wave Systems is a quantum computer company that has partnered with Google and NASA in order to develop quantum computer technology. D-Wave critics have argued that the machine is not truly a quantum computer model; supporters of D-Wave have countered that D-Wave is a quantum computer because the system runs on quantum annealing. D-Wave cannot be used for a broad range of computer applications. The quantum annealing features of D-Wave limit the machine's usage to optimization puzzles; an optimization puzzle for which D-Wave is designed to solve for the user is a scenario in which a certain destination may be reached by a very large amount of paths, and one of those paths is the fastest route to the destination. D-Wave is able to analyze the many different routes to a particular destination and determine which way is the most optimal one to take; D-Wave uses machine learning principles in order to optimize performance by determining the fastest path to take in order to reach a destination (Metz 2015). Therefore, a chief limitation to the functionality of D-Wave is the very limiting reality that the machine has been designed to specifically handle optimization problems; nonetheless, if D-Wave is capable of performing quantum speed-up, then such an ability could certainly be applied to the field of predictive analytics.

Optimization problem solving capabilities may be applicable to problem solving that is inherent to the development of AI. There is a great deal of other considerations and capabilities that would have to be considered before true AI can be invented; according to Metz (2015), classical computation utilizes binary language that reads in ones and zeroes; when the switch is yes, then a one is represented, but when a switch is no, then a zero is represented. Quantum computers are the same as classical computers in regard to binary language, but quantum computers use quantum bits, or simply qubits. Binary language processing with the use of qubits functions in a way that while a qubit is in a superposition, the quantum bit is both a one and a zero. Qubits being used in binary language allows for four times the amount of information being saved on the computer during the superposition state of the qubits, and when the state of de-coherence takes place, the qubit will then only represent one binary state which can be read as a concrete number by the CPU (Metz 2015). Therefore, the manipulation of quantum particles may allow binary language computing to achieve quantum speed-up in a way that classical computer technology never could manage to perform. Quantum computers such as the D-Wave could perform binary based functions in order to carry out calculations much more effectively than classical computing because of the fact that super-positional states allow for more work to be performed at the same time.

The designers of D-Wave may or may not have chosen the best method of attempting quantum speed-up by way of qubit operations; according to Metz (2015), the D-Wave does not use qubits as semi-conductors; rather, the qubits function as superconductor based transistors. D-Wave qubit transistor functions in cold temperature and constantly sends energy in two different directions that form a ring; algorithms are run via the superconducting qubit rings in order to calculate the answers to optimization problems. Modern computer technology is designed in a way that utilizes processors without using up very much energy, but D-Wave does not consume very much energy, and is thus able to consume energy without constraints that are related to traditional CPU functions; most of the energy consumption considerations associated with D-Wave is in regard to the machine's cooling system. The D-Wave System is potentially faster than traditional computer technology, but D-Wave's performance has been somewhat of a disappointment thus far; D-Wave has not exhibited the incredible processing speeds that experts had anticipated observing from the machine (Metz 2015). The utilization of super-conductors as a form of qubit has been the mainstay of the D-Wave; such a design has allowed for quantum computation to take place via the phenomenon known as quantum annealing. Critics may be correct in terms of arguing that super-conducting qubits that use quantum annealing is not the best method to achieve quantum speed-up; however, the true performance of the D-Wave can only be learned through vigorous testing.

Quantum speed-up cannot be achieved if the computer in question is not really a quantum machine. The designers of D-Wave can now address any critics on that particular question concerning the D-Wave and the super-conductor design; according to Brandom (2014), Google's D-Wave laboratory discovered that the D-Wave II system was running in the capacity of a genuine quantum computer, and that discovery bolstered D-Wave Systems' marketing claims that the machine's functions were in fact quantum based (Brandom 2014). The D-Wave developers have been able to silence critics that had been alleging that D-Wave was merely a pseudo-quantum machine. The use of quantum annealing through the running of super-conducting technology is a valid method of qubit design; however, the question as to whether the D-Wave is able to achieve quantum speed-up is another issue to be addressed.

The comparison of the single D-Wave computer against multiple computers working in unison as a cluster is a method of discovering if any quantum speed-up is exhibited by D-Wave over the classical processor technology; according to Brandom (2014), D-Wave, which sells at fifteen million dollars per computer, exhibited somewhat of a disappointing performance during trials. Microsoft researchers worked in collaboration with Google in order to test D-Wave against classical computer technology; D-Wave was tested in the performance field in which the quantum machine is designed to hold a comparative advantage. D-Wave was tested in terms of performance against Microsoft Research's top of the market classical computer clusters, and the results showed that D-Wave displayed no obvious performance advantage against the classical computational technology used in the trials. Quantum computers such as D-Wave are under development in order to achieve quantum speed-ups, which give the computer notably enhanced speeds, which can be of assistance to users that are running large amounts of data to be processed. D-Wave's quantum annealing functionality may or may not be able to achieve significant quantum-speedups; while the speed tests between D-Wave and Microsoft's classical clusters did not in any way show that D-Wave was not exhibiting quantum speed-ups, any quantum speed-ups may have been offered by D-wave during the tests would have been minimal; because the D-Wave System did not offer exponentially faster processing speed through quantum annealing, the high cost of creation for D-wave computers may not be financially practical. The D-Wave trials did not show that D-wave was in any way inherently prevented from faster speeds than had been shown in the tests; in fact, the pioneering D-Wave computer that was tested against Microsoft's top of the market computer clusters proved to be as fast as those more commercially developed products (Brandom 2014). The test that entailed a comparison between D-Wave and Microsoft's classical computer clusters does not seem to indicate any real signs of significant quantum speed-up; however, D-Wave has proven to be an advanced computer model. The D-Wave has already been developed and invested in heavily, and the option to dramatically shift away from the use of quantum annealing towards a different form of quantum computer method would most likely not be an option for D-Wave's developers. The uncertainty does remain over whether or not the D-Wave can actually achieve quantum speed-up, the developers of that particular quantum machine may certainly want to go back to the drawing board and find out if the D-Wave can be better designed and specified in order to perform better in any similar tests. Quantum annealing may be a useful method of quantum computing, but at the early stages of the D-Wave's existence, there does remain some serious doubt as to the real possibility of significant quantum speed-ups via quantum annealing.

    The tests have not necessarily disproven the possibility that D-Wave is able to exhibit quantum speed-up over classical computer technology. Detractors of quantum annealing being utilized in a quantum computer could argue that if such speed-up is even performed at all by D-

Wave, the advantages are minimal and difficult to observe; according to Ronnow, Wang, Job, Boixo, Isakov, Wecker, Martinis, Lidar, and Troyer (2014), the test that compared D-Wave against the performance of classical computer technology in the year of 2014, indicates that the D-Wave failed to perform faster than the classical computer technology used in the comparison tests. The model of quantum computer used in the tests was the D-Wave II, which runs on slightly over five-hundred qubits. In terms of the overall performance of computational speeds for the D-Wave during the tests, the quantum machine failed to outperform the classical computer technology; however, some controversy exists in regard to whether or not the D-Wave was faster than the classical computer technology in terms of specific periods of time during the speed trials (Ronnow, Wang, Job, Boixo, Isakov, Wecker, Martinis, Lidar, and Troyer 2014). Quantum speed-up may be very difficult to observe and appreciate on many occasions; this problem should not be a surprise to test teams, since the phenomenon that takes place in the realm of quantum physics are often very mysterious and may be difficult to associate with empirically tested computer speeds. The researchers behind the D-Wave test are able to safely declare that the overall performance of the D-Wave was not better at solving optimization problems than the classical computer clusters; however, the D-Wave managed to match those classical clusters in terms of speed.

    The designers of D-Wave decided that certain improvements could be made on the quantum machine; and such advances led to a new model. D-Wave developers were not necessarily unable to perform new tests after the disappointing showing from D-Wave against Microsoft's classical computer clusters; according to Denchev, Sergio, Boixo, Isakov, Ding, Babbush, Amelyanskiy, Martinis, and Neven (2016), processing speeds can be enhanced through the use of a quantum annealing phenomenon known as finite range tunneling. The D-Wave Two-

X has demonstrated an ability to solve tunneling problems that are related to walls of energy; this type of computational problem can also be simulated on classical CPUs; the D-Wave Two-X has exhibited higher performance during the comparison tests between classical computer technology and the new version of the quantum machine. A core processing unit ran simulated tunneling tests during a comparison trial to D-Wave Two-X, with both sides of the competition running problems using heuristic techniques in order to determine the best way to tunnel through the energy walls in the given problems. The D-Wave Two-X has been proven in the tests to be more efficient than the classical computer technology that was used as a technological rival in the recent speed trials. The tests of the D-Wave Two-X involved an algorithm known as Quantum Monte Carlo, which can be run in a simulated manner on classical computer technology. The D-Wave was found to be faster than the classical computer technology when the Quantum Monte Carlo algorithm was used in the tests; however, the researchers acknowledge that different algorithms can be used by classical computer technology that allows the non-quantum CPUs to match the performance of the D-Wave Two-X (Denchev, Sergio, Boixo, Isakov, Ding, Babbush, Amelyanskiy, Martinis, and Neven 2016). Therefore, the D-Wave performed better after the upgrade to the D-Wave Two-X; however, some validity claims could potentially be raised over the tests results. The D-Wave developers may have decided to specifically run the Quantum Monte Carlo algorithm in the test in order to ensure that the new model of D-Wave had the best chances of out-performing the classical computer technology.

Case Study Two: Test Group:

The D-Wave is not the only functioning quantum computer that exists in the world today. A team of computer scientists have invented a quantum computer that works on a qubit design that is different than the super conducting method utilized by the D-Wave. Quantum speed-up was not the subject interest when the new quantum computer was tested; according to the source titled *First Demonstration of Artificial Intelligence on a Quantum Computer* (2014), the qubits in the quantum computer exist by way of a molecular structure that is bonded to fluorine, iodine, and carbon, with one of the carbons being a number thirteen isotope. The computer's quantum bits function by being immersed in magnetic energy so that a synchronized spin can be established for the purpose of creating binary language, and the spin can be manipulated to go the other direction by means of radio wave. Zhaokai Li's team of computer scientists are even capable of manipulating each nucleus that exists inside the quantum bit molecule by means of specifically attuned wave frequencies that are intended to influence only one nucleus in the quantum bit molecule; with the concert of combined spins that take place between the nuclei allow the molecule to function in the capacity of a quantum logic gate. The one carbon atom that is a thirteen is observed by the scientists in order to ascertain if the computer has determined a character image to be a six or a nine; a spike upward indicates a six and a spike going downward shows that the computer has determined the character to represent a nine. Li's team has managed to invent a quantum computer that adequately determines the difference between the character six and the character nine, and the computer is capable to correctly perform that task even if the characters are hand written; the quantum computer's capability to perform machine learning in the character test is believed by Li and his team to be applicable to the processing of Big Data (*First Demonstration of Artificial Intelligence on a Quantum Computer* 2014). The quantum computer that was designed by Li's team may in fact be the most advanced quantum computer that has yet been created. The design may be very advanced, but the capabilities of a computer that only functions on a very small amount of quantum bits may be inherently limited. The running of Li's computer in order to exercise tasks that require artificial level capabilities is perhaps the best method to test the new quantum computer's learning capabilities.

The benefits of quantum bit design can be readily understood when the results of tests are taken into account. The objective during the tests of the quantum computer was to determine if machine learning was taking place during the time that the machine was processing information; according to Li, Liu, Xu, and Du (2015), the character for six was categorized as positive, and the character for the number nine was characterized as negative during the conditioning of the four-qubit computer during the preliminary phase of the recognition test. Handwritten characters were input into the quantum machine in a capricious manner in order to ascertain the performance capabilities of the SVM and the associated feature vectors; which are expressed in mathematical equations that relate to the numerical images of six and nine. The four-qubit computer functions by running quantum particles that function via rotating at the kernel level of the computer's circuit. The first two qubits are utilized in order to find a matrix, and during that process, the training information is recorded before the kernel matrix is established to be the same as the beginning qubit's density. The four qubit computer is then able to classify through the use of gates. The four-qubit machine's optical character recognition is made known to researchers by way of a spike that is noticeable in the computers information based spectrum, which allows the computer's users to understand that a positive classification and recognition of a numerical character on the part of the computer has occurred (Li, Liu, Xu, and Du 2015). The test results strongly indicate that some rudimentary machine learning was taking place during the time that the four qubit quantum computer was operating. The ability to differentiate between a six and a nine is not a particularly complex task for a human being; however, the fact that a computer would be able to determine the difference between two characters that happen to be hand written is a significant achievement, and shows that Li's team has invented a useful and working quantum computer.

The design of the quantum computer relies on a step process; the first inputs of information will have an impact on how the computer handles new inputs of information, and in this way, the computer is able to properly function; according to the source titled *First Demonstration of Artificial Intelligence on a Quantum Computer* (2014), the computer scientists in the team led by Zhaokai Li are affiliated with the University of Science and Technology, and the quantum computer that Li's team created is able to recognize handwritten characters through comparing the handwriting to a template. The template of characters that the quantum computer uses is constructed by images being input into a scanner, at which point the quantum computer calculates the pixel concentration of the input image so as to record a vector for the image of a numerical character. The quantum computer is then able to mathematically calculate the differences between the pixel concentrations between the character that represents six and the character that represents nine; the mathematical distinction between the six and nine characters in the computer's recorded template (*First Demonstration of Artificial Intelligence on a Quantum Computer* 2014). Therefore, machine learning is able to be performed on the four quantum bit based computer. The foundations of machine learning are laid down when the original information is input into the computer's database; then, the computer is able to make determinations in regard to any new information that is similar to the old information that already exists inside the computer's database. The process of taking in original information and then using that information to differentiate between different pieces of new information is also performed by the human brain; the quantum computer that Li's team invented appears to be capable of exhibiting a very basic form of artificial intelligence.

The difficult part of the distinction process between the characters is when the original template is relied upon to make computer judgment calls regarding new characters that are written by hand. Hand written characters may be harder for a computer to understand; according to *First Demonstration of Artificial Intelligence on a Quantum Computer* (2014), original inputs and subsequent recording of numerical characters allow the quantum computer to then make further distinctions between the characters six and nine during the part in the test that entails new inputs being scanned into and then analyzed by the quantum computer. The quantum computer is able to divide new images of the characters for six and nine, which unlike the characters used in the template, are hand written; the distinction between the two different numerical characters is achieved via the recorded numerical calculations of pixels in each of the original images that were input in the quantum computer. The team that invented to character reading quantum computer used a pioneering quantum based algorithm designed for learning in order to enable the computer to perform recognition of the images (*First Demonstration of Artificial Intelligence on a Quantum Computer* 2014). The use of a learning algorithm that is capable of being used in a four-qubit quantum computer clearly indicates that Li's team has managed to invent a very dynamic machine. The current era in quantum computer science may be very young, but the achievements that have thus far taken place are highly encouraging. Quantum computers are currently able to learn and adapt to new information, and there does not appear to be any reason why even more significant advances would not take place in the field of quantum computing. Greater investment in the development of quantum computers, both in terms of financial funding and human talent would most likely bring about more advancements and useful applications; Li's team has invented a quantum computer that could likely be useful in the field of predictive analytics.

Case Study Two: Control Group:

Machine learning is not necessarily limited to the realm of quantum computing; indeed, classical computer technology may be capable of performing such tasks. Computer tests that have been conducted in recent years have examined the ability of classical computer technology to perform tasks that require some level of machine learning. An interview between Amara Angelica and SAIL's Jurgen Schmidhuber (2012), explains that the Swiss Artificial Intelligence Lab has designed computer technology that is inspired by biological neuron networking; the computer technology that has been designed by the Swiss team has performed well in computer imagery recognition tests because of the technology's ability to perform machine learning capabilities. The Swiss Artificial Intelligence Lab won a the ICDAR Offline Chinese Handwriting Competition in the year of 2011; the Swiss designers of the winning program were not literate in the Chinese language that was used in the competition, but the recognition program that the team had entered into the competition became the winner due to superior performance (Angelica and Schmidhuber 2012). The computer technology created by the Swiss Artificial Intelligence Lab is a testament to the effectiveness of neuronal computer designs. Classical computer technology is capable of being designed to learn because of the new techniques that now exist in computer science. Quantum computing is not the only form of computer technology that is currently capable of performing machine learning functions.

The four qubit computer that was created by Li's team of computer scientists could be compared to the technology that has been invented by the Swiss Artificial Intelligence Lab. Similarities between the two different computer technologies may be found in terms of the methods by which the machines learn; according to Angelica and Schmidhuber (2012), the CPU's functional design used in the pattern recognition program that the Swiss Artificial

Intelligence Lab invented was different than more mainstream computer program models. The image pattern recognition program designed by the Swiss team utilized graphic cards in order to exponentially improve the accuracy of the program's performance. The memory of the neuronal inspired program works on a both long-term and short-term basis, and manages to calculate the probabilities that previous patterns will be encountered in the future (Angelica and Schmidhuber 2012). The computer technology that has been made from the Swiss Artificial Intelligence Lab and the four qubit quantum machine that was designed by Li's team both rely upon the intake and analysis of original information in order to make determinations about new information inputs. The computer award winning technology of the Swiss Artificial Intelligence Lab may only utilize classical computer technology, but the fundamental machine learning philosophy is similar to Li's quantum computer. Human beings learn from taking in information that will be analyzed in order to make sense of new incoming information, and the neuronal inspired design of the Swiss Artificial Intelligence Lab appears to have invented computers to learn in a similar manner.

    The comparison between the learning style of the human brain and the computer technology designed by the Swiss Artificial Intelligence Lab (SAIL) was part of a performance test of one of the Swiss lab's products. The straight forward method of simply examining the performance levels of a computer against the performance levels of a human being was used in the test; according to Angelica and Schmidhuber (2012), the innovative design of the Swiss team's program was instrumental in their technology's winning performance in the Traffic Sign Recognition Competition. The tests involved pattern recognition tests related to traffic signs and was the eighth pattern recognition competition that the Swiss Artificial Intelligence Lab had won, and demonstrated that the team's program could outperform human beings in the task of pattern recognition (Angelica and Schmidhuber 2012). The ability for the SAIL's computer design to perform machine learning allowed that particular model to best human competition in the test; however, the computer model is most likely not as versatile as a human brain, and cannot learn from a diversity of information in the same way as a human brain is capable of learning. The SAIL's computer that won the traffic sign competition was very good at learning from traffic sign patterns, but the human brain may be better at learning from a broader range of information, such as patterns found in classical music or facial expressions. The SAIL's computer technology is very good at learning information that is strictly related to specific tasks, and appears to be superior to human brains in terms of learning on some occasions.

The task of recognizing images and then being capable of learning patterns that exist between the images entails the ability to learn. Human beings are able to learn patterns that exist between images because of the way that the human brain functions; computer scientists are able to learn from the efficiency of the brain in order to design computers that may be able to recognize relationships between images; according to Angelica and Schmidhuber (2012), the Swiss Artificial Intelligence Lab has designed computer technology that is inspired by biological neuron networking. SAIL's computer technology has performed well in computer imagery recognition tests because of the technology's ability to perform machine learning capabilities. The Swiss Artificial Intelligence Lab managed to design a program that bested any of the computer programs and human rivals in a pattern recognition contest; the program designed by the Swiss Artificial Intelligence Lab was over ninety-nine percent accurate during the tests (Angelica and Schmidhuber 2012). The performance abilities of SAIL's computer technology is very impressive; the achievement of out-performing other computer competitors in the competition was clearly an indication of SAIL's innovative and superior computer design. The computer that SAIL designed for the Traffic Sign Pattern Recognition Test was superior to the other computers that performed in the competition; indeed, the model by SAIL proved to be superior to the human competition that had competed in the traffic sign pattern recognition test. The award winning computer technology that has been made by SAIL has been able to soundly demonstrate the ability of computer technology to perform machine learning functions.

The ability for a computer to recognize patterns related to images appears to now have reached the performance levels of a human being's brain, or perhaps have slightly surpassed some abilities which the human brain is capable of performing. The reasons behind why SAIL's computer is able to outperform other computer technology and even match or exceed human competition in the specific pattern recognition test is most likely because of SAIL's new and innovative design, which was invented to function in a manner similar to the neuronal firings of the human brain; according to Markoff (2012), the Swiss Artificial Intelligence Lab's technology competed in a contest in order to recognize patterns of road signs. The computer program made by the Swiss team managed to successfully identify patterns found in fifty-thousand images with over ninety-nine percent accuracy; the best human that took part in the same competition also managed to recognize patterns with just over ninety-nine percent accuracy. The Swiss Artificial Intelligence Lab designed the computer technology that won the image pattern recognition contest in a manner that emulates a brain, and this allows for machine powered deep learning to take place (Markoff 2012). The human brain's neuronal functions appear to be a very effective model to mimic if computer scientists intend to further advance the abilities of computers to perform machine learning tasks. The computer technology appears to be capable of performing machine learning in a manner that is not entirely dissimilar to the four qubit computer that was designed by Zhaokai Li's team; however, certain factors need to be taken into consideration, both Zhaokai Li's four-qubit computer and SAIL's neuronal classical computer are examples of newly emerging computer technology. The four qubit computer is an example of quantum machine learning, and the SAIL's computer technology is an example of neuronal computing, and the fact that both computer designs are able to learn is an indication that quantum computing and neuronal computing are able to perform machine learning.

The examination and analysis of the case studies indicate that the independent variable has a reasonably strong impact upon the dependent variables; in other words, the operationalization of the variables shows that the independent variable positively influences the dependent variables. The results found after the operationalization of the variables supports the hypothesis; however, the impact of the independent variable was somewhat moderate, and therefore, limits the scope of the hypothesis's expectations. The hypothesis is found to be supported, but certain inherent limitations are found to exist concerning quantum computer technology's advancement of computer driven predictive analytics.

The D-Wave's quantum computational ability to solve optimization problems is at least as efficient as the most advanced classical computer technology; the fact that a single computer running on merely roughly five hundred quantum bits could match the processing power of entire classical computer clusters is a very strong indication that quantum computing technology is far superior in certain respects over classical processing models. Classical computer technology has been under incrementally advancing development for decades, and computer engineers have become very skilled at designing such technology; quantum computer technology on the other hand, has only recently emerged beyond prototype stages of development. The fact that a relatively recently developed quantum computer model such as D-Wave could match the performance of some of the best classical computer clusters clearly shows that quantum computer technology is the next step in computer evolution.

The optimization problems that D-Wave has been invented to solve can be applied to the intelligence field of predictive analytics. The optimization answers that the D-Wave is capable of providing after analyzing large amounts of information would be very useful when analyzing Big Data; however, because of the fact that the D-Wave is limited in scope to only be able to solve certain types of optimization problems, that particular quantum computer is incapable of solving a variety of Big Data problems that do not pertain to optimization problems. The diversity of Big Data that must be analyzed in the field of predictive analytics is very large, and the D-Wave is only able to solve problems for a narrow spectrum of information found in Big Data. Superconducting qubits that utilize the phenomenon of quantum annealing may only be able to perform certain functions; that being considered, some basic AI functionality may be offered by the D-Wave computer design. The ability of the D-Wave to perform optimization solutions entails that the quantum computer possesses some rudimentary problem solving skills, which is a basic component of artificial intelligence.

New versions of the D-Wave that would run on more qubits than the current models, and are more sophisticated utilization of machine learning algorithms would most likely lead to future version of the D-Wave being able to continue to achieve even higher levels of processing power. In the future, better versions of D-Wave may periodically be developed; thus, Moore's Law would continue to be proven correct. The possibility of ever advancing versions of D-Wave would offer quantum computer technology the ability to increasingly perform at the level of artificial intelligence; likewise, such AI level capabilities would give future models of D-Wave a place in the field of predictive analytics that match or exceed the problem solving abilities of human beings in certain areas of analysis. New models of D-Wave would be able to match or even exceed human beings in certain areas of analysis that is applicable to predictive analytics; nonetheless, more advanced D-Wave models would be inherently limited to specific roles as an artificially intelligent analyst. Human beings would still be able to outperform future D-Wave models in a wide range of analytical tasks that pertain to the field of predictive analytics; therefore, both human analysts and future models of D-Wave would benefit by the synergy that would come through the combined effort of analyzing information in order to produce predictive analytics reports.

The four qubit quantum computer that was invented and tested by the team led by Zhaokai Li appears to be capable of machine learning; the computer is able to analyze imagery input data and then take into account that inputs relevant details in order to make determinations about new but similar inputs, and in doing so is able to perform a machine learning task. The neuronal inspired computer technology that has been designed by the SAIL is classical computer technology, and is not a quantum computer; however, the SAIL's neuronal computer design is able to learn from previous imagery inputs in a manner that is not dissimilar to Zhaokai Li's quantum computer.

The machine learning capabilities that have been displayed by the SAIL's classical but neuronal inspired computer technology and Zhaokai Li's four qubit quantum computer both display artificial intelligence related skills that are applicable to the field of predictive analytics; indeed, Big Data that is often relevant to the field of predictive analytics is many times in the form of imagery. The ability for computers to learn from previous images that exist in Big Data and then make determinations about new but similar data would allow for more sophisticated predictive reports being produced. The quantum computer designed by Zhaokai Li's team and the neuronal computer technology designed by the SAIL could feasibly be combined into a form of computer network or cluster in the future, and this combination of technology would allow for an increased level of artificial intelligence. The ability for computer technology to associate relationships of similar imagery that exists in a sea of Big Data, and then further make determinations in regard to such imagery by recognizing patterns would greatly assist the effort to produce predictive reports. Quantum computer technology that is capable of machine learning and happens to operate on an advanced neuronal design would essentially be an artificially intelligent analyst; however, the case studies indicate that even after a fusion of the four-qubit quantum computer invented by Li's team and the SAIL's neuronal technology, would together, only be capable of analyzing certain forms of information found in Big Data. The diversity of information that exists in Big Data is incredibly diverse, and analyzing imagery is only one specific skill-set that can be utilized in the effort to produce predictive reports based on intelligence derived from Big Data. Quantum and neuronal computer designs may someday be capable of performing on or above the same performance levels as is routinely found among human analysts; but, human analysts would continue to be a necessary component in the predictive analytics field because the human intellect would still possess broader capabilities than future models of quantum computer technology.

The research question is able to be answered based on the findings reached from the operationalization of the variables. Quantum computer technology could someday allow computer technology to perform machine learning and problem solving skills, and such abilities would allow quantum computer technology to independently analyze Big Data in order to produce predictive analytics reports; likewise, such computer advances would need to be combined with the skills of human analysts in order to produce predictive reports that comprehensively include analysis derived from a broad range of information. Quantum computer technology will most likely not be able to replace human beings in the field of predictive analytics, but the further development of quantum computers will allow cyber-technology to evolve from mere tools used by human analysts, to artificially intelligent analysts that would work alongside human beings in the intelligence field of predictive analysts.

Conclusions:

The research results suggest that quantum computers will allow for Moore's Law to continue being correct, and computer technology would incrementally advance at a steady rate. Future models of quantum computers could act in an unsupervised or semi-autonomous manner because of AI level functionality, such as machine learning and problem solving skills. Certain aspects of the field of predictive analytics will possibly be handled almost exclusively by quantum computer technology because of comparative advantages that such future artificial intelligence would most likely be capable of exercising. Specialized quantum computer technology would most likely be used to produce sophisticated and high quality analysis, but human analysts would be irreplaceable during the subsequent steps of the intelligence process. Human analysts would be needed to further analyze quantum computer outputs that concern answers found in Big Data; analytical reports that human analysts produce would most likely be combined with the reports that were produced by the specialized quantum computer technology, and comprehensive predictive reports would then be made available to the intelligence consumers. Human analytic abilities being someday combined with future specialized quantum computer technology would most likely be a profoundly beneficial step forward in the evolution of the intelligence field known as predictive analytics; ultimately, the research finds that quantum computer technology could be combined with other computer science breakthroughs, such as neuronal computational networking, and such cyber-technology would most likely allow for artificially intelligent computers to solve problems and learn while aiding humans in the overall predictive analytics process.

Some experts in the field of artificial intelligence have put forward claims that the human brain is far too complex and sophisticated to be matched in cognitive functions by computers; such skeptics of the potential of artificial intelligence have argued that computer technology can essentially play the role of little to nothing more than a tool that human analysts can utilize in the field of predictive analytics. A school of thought among experts that work infields related to artificial intelligence has argued that computer technology can be invented or has recently been invented that can match the analytical skills of human beings; such experts believe that artificial intelligence is possible, and that such machines can be given the responsibility of offering answers based on computer powered analysis. The debate over the potential of artificial intelligence continues, but some scientists have argued that the rise of quantum computer technology may offer the metaphorical key to developing artificial intelligence that could match or exceed the performance of human analysts; such capabilities may someday exist because of the problem solving and machine learning skills that quantum computing technology may eventually enable computers to exercise. The research set out to answer the following *research question*: in what ways can artificial intelligence based predictive analytics be advanced by quantum computing?

In the spirit of finding answers to the research question, a methodology was devised which entailed testing the following hypothesis: despite computer based predictive analytics technology currently being only capable of producing reports that are of a quality which lack the analytical problem solving capacities of human analysts, the development and subsequent applications of quantum computing technology will bring computer driven predictive analytics to at least the same level of analytical problem solving capabilities as is routinely offered by human analysts.

Case studies were selected in order to operationalize independent and dependent variables. Control group case studies examined advanced classical computer technology. The research's test groups examined computer problem solving skills, produced probability outputs, and machine learning capabilities of quantum computer technology that can be used in the field of predictive analytics. Qualitative and quantitative measurements were considered in order to fulfill the operationalization of the research's variables.

The <u>independent variable</u> selected in the research was quantum computer technology that is being applied or is believed possible to be applied to AI designs that could be used in the field of predictive analytics. The <u>dependent variables</u> selected in the research were computer based problem solving capabilities, probability outputs, and machine learning capabilities that can be applied to the field of predictive analytics. The criteria for understanding the research's findings after the operationalization of the variables via analysis of the selected case studies was set forth in the following way: if the independent variable significantly impacts the dependent variables examined in the research, then the hypothesis will be found as supported. However, if the impact of the independent variable upon the dependent variables is found to be insignificant/negligible, then the hypothesis used in the proposed research will be found to be unsupported. The results that were found after the operationalization of the research's variables show that the hypothesis was moderately supported.

The research offers several suggestions for important future research that relates to the subject matter that this research has examined. Future research should be carried out in regard to how predictive products such as Watson Analytics can be improved upon by the incorporation of quantum computer technology; quantum computer technology is currently available in different design models, and uncertainties in regard to the exact D-Wave's benefits to the field should be addressed in future research. Future research in regard to the benefits of incorporating the D-Wave design into the field of predictive analytics could possibly attempt to find answers by approaching the usefulness of integrating technology that is specifically intended to solve optimization problems to work in conjunction with the established cyber-technology that is currently used in the field of predictive analytics. Greater efforts towards learning how to fully and effectively integrate a quantum computer such as the D-Wave design would be useful; indeed, such research would most likely not only be insightful from the point of view of scholars and intelligence professionals, such future research would also offer the benefit of allowing the D-Wave developers a clearer view of the marketability of the D-Wave as a product in the field of predictive analytics.

Future research may also be directed towards the unanswered question of how exactly any future models based on Zhaokai Li's four qubit quantum computer design could be integrated to work alongside classical predictive analytical technology such as Watson Analytics. The quantum computer designed by Zhaokai Li's team may offer significant benefits to the field of predictive analytics. People that are familiar with the characteristics of social media communication would almost certainly be aware that imagery and words are often times placed together in order to convey meanings that often times trend over the Internet, and a computer that is capable of image related machine learning would offer possible benefits if incorporated into predictive products such as Watson Analytics. Future research that could answer the questions that relate to how best to apply quantum computer models would be very helpful to many parties that practice or seek to improve the performance capabilities of predictive analytics cyber-technology.

Future research may also wish to focus on understanding how newly emerging classical computer technology such as the program known as Eugene could be designed to work alongside quantum technology in the form of integrated computer networks that combine some quantum computer technology with newly emerging classical computers with advanced abilities to understand human conversations. Products which are currently being marketed for use in predictive analytics may be improved by the application of programs that are designed to understand human conversation; many conversations take place over the Internet such as communications that happen every day via social media, and such data is often available to be analyzed inside a sea of Big Data, therefore, products such as Watson may be improved if programs such as Eugene are integrated into that product. Quantum computer technology that may be used in the predictive analytics field would need to be integrated alongside such technology, and future research should examine how such integration could be best achieved.

Research that is conducted as in order to follow up on this research may want to focus on neuronal inspired computer designs as a possible next step in the evolution of computer technology, and how such neuronal designs could someday be invented to work as computer clusters that include quantum computing technology. Future research could also examine whether or not neuronal inspired classical computer technology could be invented to work inside a quantum computer as a single model. The question of how the two emerging cyber-technologies could be somehow combined in order to bring about a synergy of the efficiency of artificial neurons and quantum speed-up would be a very worthy course of future research. Ultimately, incredible discoveries may be the result of future research that endeavors to further understand the true workings of the human brain, and any quantum manipulations that the brain may be utilizing in order to bring about cognitive functions. Artificially intelligent quantum computers that someday exercise the problem solving and machine learning skills needed to produce predictive analytics reports in a manner that is unsupervised by human beings may possibly only be invented if researchers first learn how the human brain's functions can be applied to computer technology.

References:

Albert, Hans, Michelle Simmons, and Yuri Samoilov. "Quantum Computing." *Nova*, February 18, 2016. Accessed February 18, 2016. http://www.nova.org.au/technology-future/quantum-computing

Angelica, Amara, and Jurgen Schmidhuber. "How Bio-inspired Deep Learning Keeps Winning Competitions." *Kurzweil*, November 28, 2012. Accessed February 13, 2016 http://www.kurzweilai.net/how-bio-inspired-deep-learning-keeps-winning-competitions

Brandom, Russell. "Google's Quantum Computer Just Flunked Its First Big Test." *The Verge*, June 19, 2014. Accessed February 06, 2016. http://www.theverge.com/2014/6/19/5824336/google-s-quantum-computer-just-flunked-its-first-big-test

Charniak, Eugene. "Bayesian Networks without Tears," *AI Magazine, 12* (4). (1991): Accessed December 11, 2015 http://www.aaai.org/ojs/index.php/aimagazine/article/view/918/836

Chomsky, Noam and Nikola Danaylov. "Noam Chomsky: The Singularity is Science Fiction." *Singularity 101 Weblog.* October 4, 2013. Accessed December 11, 2015 http://www.singularityweblog.com/noam-chomsky-the-singularity-is-science-fiction/

Cuthbertson, Anthony. "IBM Watson CTO: Computing Could Advance Artificial Intelligence by Order of Magnitude," *International Business Times,* 2015, Accessed December 11, 2015 http://www.ibtimes.co.uk/ibm-watson-cto-quantum-computing-could-advance-artificial-intelligence-by-orders-magnitude-1509066

Denchev, Vasil, Sergio Boixo, Sergei Isakov, Nan Ding, Ryan Babbush, Vadim Smelyanskiy, John Martinis, and Hartmut Neven. "What Is the Computational Value of Finite Range Tunneling?" *Quantum Physics, Cornell University Library*, January 22, 2016. Accessed February 15, 2016. http://arxiv.org/abs/1512.02206

Domingos, Pedro. "A Few Useful Things to Know about Machine Learning." No date. Accessed January 12, 2016. https://homes.cs.washington.edu/~pedrod/papers/cacm12.pdf

Fedichkin, L.E., Yanchenko, M.V. and K.A. Valiev. "Novel Coherent Quantum Bit Using spatial Quantization Levels in Semiconductor Quantum Dot." (2000): *Institute of Computer Science.* Accessed December 11, 2015 ics.org.ru/eng?menu=mi_pubs&abstract=249

Author, No. "First Demonstration of Artificial Intelligence on a Quantum Computer." *The Physics ArXiv Blog*, October 14, 2016. Accessed February 20, 2016. https://medium.com/the-physics-arxiv-blog/first-demonstration-of-artificial-intelligence-on-a-quantum-computer-17a6b9d1c5fb#.yp6a2ojb7

No author. "How Does Predictive Analytics Differ from Data Mining and Business

Intelligence?" *FICO DM Blog.* Last modified 2006. Accessed December 11, 2015 http://dmblog.fico.com/2006/06/how_does_predic_2.html

Hsu, Jeremy. *Google Tests First Error Correction in Quantum Computing.* IEEE Spectrum, 2015. Accessed January 14, 2016 http://spectrum.ieee.org/tech-talk/computing/hardware/google-tests-first-error-correction-in-quantum-computing

Li, Zaokai, Xiaomei Liu, Nanyang Xu, and Jiangfeng Du. "Experimental Realization of a Quantum Support Vector Machine." *APS Physics, Physical Review Letters*, no. 114 (April 08, 2015). Accessed February 13, 2016. http://journals.aps.org/prl/abstract/10.1103/PhysRevLett.114.140504

Lohr, Steve. "Creating Artificial Intelligence Bases on the Real Thing," *The New York Times,* 2011, accessed December 11, 2015 http://www.nytimes.com/2011/12/06/science/creating-artificial-intelligence-based-on-the-real-thing.html

Lloyd, Seth, Masoud Mohseni, and Patrick Rebentrost. "Quantum Algorithms for Supervised and Unsupervised Machine Learning." 2013. http://arxiv.org/pdf/1307.0471v3.pdf

No author. *Machine Learning.* No date. Accessed January 12, 2016. http://www.sas.com/en_us/insights/analytics/machine-learning.html

Markoff, John. "Scientists See Promise in Deep-Learning Programs." *The New York Times*, November 23, 2012. Accessed February 16, 2016. http://www.nytimes.com/2012/11/24/science/scientists-see-advances-in-deep-learning-a-part-of-artificial-intelligence.html?_r=0

Mengshoel, Ole. 2010. "Understanding the Scalability of Bayesian Network Interface using Clique Tree Growth Curves." *Artificial Intelligence.* (2010): Accessed December 11, 2015 http://ac.els-cdn.com/S0004370210000846/1-s2.0-S0004370210000846-main.pdf?_tid=67fdfc54-a0a3-11e5-a8cc-00000aacb361&acdnat=1449906119_200eeba0894bbc5707ff21a245c818cf

Metz, Cade. "Google's Quantum Computer Just Got a Big Upgrade." *Wired*, September 28, 2015. Accessed February 06, 2016. http://www.wired.com/2015/09/googles-quantum-computer-just-got-a-big-upgrade-1000-qubits/

Rigatos, G.G., and S.G. Tzafestas. *Quantum Learning for Neural Associative Memories* 157, no. 13 (2006): 1797-813. Accessed January 12, 2016. http://www.sciencedirect.com/science/article/pii/S0165011406000923

Ronnow, Troels, Zhihui Wang, Joshua Job, Sergio Boixo, Sergei Isakov, David Wecker, John Martinis, Daniel Lidar, and Matthias Troyer. "Defining and Detecting Quantum Speedup." *Science*, July 25, 2014, 420. Accessed February 15, 2016. http://science.sciencemag.org/content/345/6195/420

Siegelmann, Hava and Eduardo Sontag. 1991. "Turing Computability with Neural Nets." *Rutgers University Appl. Math. Lett. Vol. 4 No. 6.* (1991): accessed December 11, 2015 www.math.rutgers.edu/~sontag/FTD_DIR/

Simmons, Michelle. "Quantum Computation." *TEDxSydney.* 2012. Accessed February 19, 2016. http://www.nova.org.au/video/quantum-computation

Author, No. "The Moore's Law of Big Data." *National Instruments*. January 15, 2013. Accessed February 20, 2016. http://www.ni.com/newsletter/51649/en/

Thompson, Clive. What is I.B.M.'s Watson? *The New York Times Magazine,* 2010, accessed December 11, 2015 http://www.nytimes.com/2010/06/20/magazine/20Computer-t.html?_r=0

No author. "Turing Test Success Marks Milestone in Computing History." 2014. Accessed January 12, 2016. http://www.reading.ac.uk/news-and-events/releases/PR583836.aspx

IBM. "What is Watson Analytics?" *IBM Watson Analytics.* Last modified November 5, 2015, Accessed December 11, 2015 www.ibm.com/analytics/watson-analytics/

Ying, Liu. "Big Data and Predictive Analytics." *The Journal of Business Forecasting* (2014): Accessed December 11, 2015 https://www.questia.com/library/journal/1P3-3601906361/big-data-and-predictive-business-analytics

Zhong, Manjin, Morgan Hedges, Rose Ahlefeldt, John Bartholomew, Sarah Beavan, Sven Wittig, Jevon Longdell, and Matthew Sellars. *Optically Addressable Nuclear Spins in a Solid with a Six-Hour Coherence Time* 517, no. 7533 (2015): 177-80. Accessed January 14, 2016. http://www.nature.com/nature/journal/v517/n7533/full/nature14025.html

www.ingramcontent.com/pod-product-compliance
Lightning Source LLC
Chambersburg PA
CBHW080537190526
45169CB00007B/2535